Thomas Laidlaw

The Old Concession Road

Thomas Laidlaw

The Old Concession Road

ISBN/EAN: 9783744679343

Printed in Europe, USA, Canada, Australia, Japan

Cover: Foto ©berggeist007 / pixelio.de

More available books at **www.hansebooks.com**

THE OLD

CONCESSION ROAD,

——BY——

THOMAS LAIDLAW.

The speckled trout, when we were boys,
That finned the shady streams,
And glanced above the sandy bars,
Are flashing through our dreams.

GUELPH:
O. E. Turnbull, Printer and Binder.
1899.

IN MEMORY OF THE EARLY SETTLERS.

In the multiplicity of books is there room for another? Even for such as this—is there?

A few years ago a very humble edition of "The Old Concession Road" was published, which in one way or other has been disposed of. The writer has been advised to have it reproduced, which he has ventured to do; but whether wisely, or otherwise, is an uncertainty. In this edition a section has been added, others have been enlarged in a greater or less degree, a picture of the old log school house is given as a frontispiece and in form the book is more attractive.

The little work pictures the early years of a settlement in which the greater part of the writer's life was spent and it is believed to be typical of others over the length and breadth of our land. It exhibits the spirit that was witnessed in the life of the early settler, it delineates the personal experiences and observations of the writer, the memory of which has for him a singular fascination and he lived through the whole.

The lyrics which follow are the same as in the former edition.

THOS. LAIDLAW.

Guelph, 1899.

IN THE DAYS OF THE LITTLE SHANTY.

To love we would the task resign
And from oblivion wrest
Scenes of the old concession line,
When first by traffic prest.

Ah! he who sings was then a boy
Bareheaded and unshod
And sees in age with chastened joy
The old Concession Road.

Recollection of how we entered the old concession
is at this date rather hazy, though we incline to think
that we were taken there in a lumber wagon drawn by
a yoke of oxen.

The mists of sixty years lie between us and that
event, therefore, many a scene of real pathos not affect-
ing the great selfish world at large, but deeply interest-
ing to the worthy settlers on the old concession road
is now lost in obscurity. Sixty years have a dimming
effect. Even the shorter catechism, that was drilled
into us sixty years ago in a way which the mothers of
that generation only knew, has in many of its lines
faded, though its starting point, "What is the chief
end of man?" is, we suppose, written with indelible
ink. A track was cut through the woods before this
to the township of Waterloo, in the direction of Berlin,
for the purpose of bringing supplies from the worthy
Dutchmen settled there, into the village of Guelph.
For John Galt, four years previous to the humble event
to which we have referred, had cut the first tree and
laid the foundation of the future city, which we
always feel to be somewhat similar to ancient Jerusa-
lem. "As the mountains are round about Jerusalem"
so are the hills around Guelph, though here the com-
parison ends, as it is not "a city that is compact

together.'' It is likely that we were taken part of the way over this track.

It was in the fall of this year (1831) that the line was opened out. O! for a pen to sketch vividly the concession road of those early years! And who would be interested in that though you had? What distinctive features have this line over many others that it should be so estimated? Are there not a hundred such elsewhere as worthy of our wonder? It is not for us to say that there are not a hundred more worthy, but then this particular one is ours and that makes all the difference. Yes, it is ours, and has been from our early youth. Over its rough uneven track we have run—a boy—barefooted and scaithless, though the sun was beating on our uncapped head; indeed our only covering a shirt and trousers, the latter kept in position by one solitary suspender. We saw it in the beginning stretching away in perspective through the grand old woods, whose tops gracefully gave edgings to a strip of stainless sky. Our life is linked with it —it is part of ourselves. To its joys and sorrows the heart vibrates as an Æolian harp to the zephyr. Reasons sufficient for the wish to spare its early scenes from the bleachings of time; to preserve the aroma of its native woods, and to gather up its dying echoes ere they are lost in forgetfulness. Our pen is unequal to the task and will assuredly fail, but, if fail it must, there shall be this satisfaction, that it did its best.

In the concession a majority of the settlers are from the Lowlands of Scotland, with the merest sprinkling of the Highland Celt, a family from the Emerald Isle, a few indicate by their tongue that they are cradled south of the Tweed, and lastly a family or two from Fatherland are tucked into the line at its western extremity. Such are they who cut the first tree and raised the little shanty in the woods, as they rise in memory before us and after a calm reflection of many years, we say it in verity—they were worthy of the several races from which they sprang. If we are partial the reason has been given, the associations of youth. We love our own, yet in a sense we are lifted above such geographical distinctions—we recognize the sisterhood of concessions. Have we not been jolted over the corduroy of many of them in an ox wagon. But better still we knew the worthy residents for years and learned to respect them, we have sat at their simple board and shared in their frugal meal, aye, and we have drank of their whiskey and called it good.

In fancy from an imaginary summit we see the whispering forest of those days stretching in every direction away to the horizon in wavy undulations. No pen has written its history save the pen of Him who is the Ancient of Days; yet in ages remote we catch in imagination the first ripple of the streams in their outward flow down the solitary valleys, a vegetation creeps over the dreary waste, forests spring up and expand, birds nestle in their branches and beasts haunt their solitudes, and then a figure lithe and agile with bow and arrow is seen gliding with stealthy step and peering with cunning eye through the interstices of the wood—centuries roll, and now in the fulness of time smoke is seen curling in wreaths from little openings in the wood and the knell of the drowsy past breaks in echoes at our feet.

THE SUGAR CAMP.

Stars twinkle from a summer sky,
And through the pale moonlight,
The sweet clear notes of whippoorwill
Enrich the silent night.
But list! the far resounding horn
Is doubling through the woods;
Some hapless one has lost his way
In pathless solitudes.

Our concession, we delight to call it ours, is now opened—a track has been cut through the woods! At intervals along the line are the settlers' houses, built with logs, simple in construction, and not a few are mere shanties, separated only by a short distance, yet each in its own little clearance is completely isolated. And it is true that a worthy settler in this very locality, with the sun shining clearly, lost his way in crossing from a neighbor's house to his own—though across the distance a voice wood be understood.

An axe is heard in the woods and in the direction of the echo an acre or two of clearance is seen with a little shanty.

Among the stumps are potatoes growing in hills, the only way in which they could be cultivated, and pumpkins with their long trailing vines ripening in the warm glowing sun, and there too is the ubiquitous sunflower, which is never missing, a genuine pioneer among flowers and in true harmony with its surroundings. Its full round honest face evokes admiration, looking at the sun and everything else with an open frankness. It may be that the door of the little house is ajar and we look in. Conscious of guilt, a barn yard fowl or two flutter hurriedly past, and it may even be from under the bed, but we would not have that repeated for the world, as very likely the worthy woman is out helping her husband to burn brush and gather up chips and roll logs, or something else, in a way which we who recognize the dignity of labor know how to appreciate.

We go up one concession and down another, taking in the entire settlement, studying the situation and character of the people. They are poor with few exceptions, but rich in habits of industry and in being inured to toil. In that humble abode with its low rough walls, around which the winter's snow is eddying and the stars are yet glistening clearly on the snow-hooded stumps that stand thick in the little clearance, the inmates have already partaken of their simple meal and he who is the humble hero of the scene lifts the latch—locks were never thought of, the moral worth of the concessioners was a surer safeguard than bars of steel—so he lifts the latch and with his axe goes out into the leafless woods. He toils till the night shades gather and deer come out from the thickets, under cover of the dusk, as they often did, to browse from the heap at which he wrought. And then the snow begins to melt, for the days are getting long, soft winds are sobbing through the trees, and the sky is interesting; there is expectancy in the air —a listening as if for something that is coming, the peebee is heard in the woods and the sugar-makers follow to where it is leading—so sugar making begins. Trees are tapped and sap gushes from the fresh wounds pure and sweet. We are alone in the bush and on this day there is a delicious softness in the air, sensitive and impressive to the faintest whisper; very audibly we hear the sap drop, drop, dropping with a peculiar sound into the little troughs all over the bush, and so continuously. No, we are not alone. There is an invisible presence around us, which we feel to be very real, and there is a weirdness which we cannot express. And the sap keeps drop, drop, dropping as if one little drop was calling to another through the enchanting stillness. No, we are not alone! And the camp fire is kindled, huge logs are rolled in on either side of the kettles and steam rises from the boiling, bubbling sap. What a day for making sugar! The season is at its best, with the sun seen as through a glass, shining with chastened light, often from behind a cloud that might drop out in rain, the wind soft and dirgeful as if nature were in grief. How the sap runs—it gushes! It is obvious that by sunset the troughs will be filled to overflowing. Sap that was boiled during the night is now strengthening to the point of sugaring off. As it gets nearer and nearer to the required consistency with what intensfied interest the youth of the encampment look on, as ever sweeter, sweeter, sweeter grows the syrup in the kettle and in what various ways it is tested and tasted to the last. Is it a dream! The merry voices of children come up to us echoing from the quiet shades of the forest and our old familiar friends are gathering in beside us as in other years till we almost feel the warm pressure of the hand that has been long missing and hear the voice that has been long still. We confess to the stirrings of tenderness. Yet never-more will the camp fire be lighted in the quiet shades of the forest, it is quenched with the council fires of the dusky aborigines and the light of the lonely wigwam.

Then spring came with opening buds and softer skies, with immense flocks of wild pigeons seemingly of unlimited extent, covering the whole face of the visible sky, winging their way through the sun-lit atmosphere. And we saw them with wonder. Seed was committed to the virgin soil, and not without hope, spring deepened into the warm leafy summer, woods

were rank with herbage, and all were busy burning
timber cut through the winter. Then the oxen were
brought in from the woods in the morning, the yoke
placed on the neck of one, the other end raised up, the
words given—"come under "—and the patient, honest
beast came up with his wine colored eyes, meekly had
the yoke lowered to his neck and made secure, and
was driven to the logging field for the day, and boys
entering their teens were in request and maidens in
the bloom of womanhood gave their services to the
work. And when the toils of the day were over and
the night settled in and around the little dwelling, in
the far off recesses of memory we see the fireflies
twinkling in the gloom, the light reflected from the
burning logs, the cows burying themselves in smoke
from the dreaded mosquitoes, and we hear, as if it
were in echoes, the whippoorwill making the skirts of
the little clearance vocal with its plaintive notes, and
the wolf's howl as it rings from the deeper solitudes,
like to a voice from behind the mountains.

Years passed away and the settler's axe still con-
tinued to ring through the snows of winter, summer
brought its own peculiar occupation. So the little
clearance grew and broadened field was joined to field,
and light broke in and across the concessions, through
the openings. The sound of the horn was no longer
heard when it was feared that some one had lost his
way in the woods hunting cows, as we ourselves
knew one, and a young girl at that, who was out all
through a rainy night, seated lonely on a fallen tree,
her ear intensely acute to the slightest sound through
fear. And the little harvest grew till the sickle had
to yield to the greater capacity of the cradle, and the
double log barn was inadequate to contain the ripened
sheaves. Then the Indian disappeared from the con-
cessions with his venison, yet ghost-like we saw him

brushing the hoar-frost from the fallen leaves with his moccasins in the early dawn, gliding stealthily through the fragmentary woods away to the setting sun, to have rest only in the happy hunting grounds of his fathers.

A SABBATH IN THE EARLY DAYS.

> A day of rest, no grating note
> Disturbs this brooding spell,
> No, voice save nature's blending with
> The tinkling cattle-bell;
> The weary prize this precious gift—
> A holy Sabbath calm,
> The reverend woods their voices lift
> And sing their hymn and psalm.

Six consecutive days in a logging field involve a great deal of grimy work, and he who is otherwise than black as an Ethiopian at the end of a week is untrue. Our settlers were not of that class, a thorough wash and a clean shirt were necessary, and then at the merest wink of their weary eyes the sleep which waits on honest toil took them to her downy breast, within their rude chinked walls. Then the morning dawned fresh and beautiful as on that day when it received its sanctified impress; and the sober looking oxen, stiff from a hard week's work, slowly raised themselves up from their leafy beds, and quietly began to crop the herbage on the edge of the clearance with a semi-consciousness of a day of rest: and the shifty little cows were on the move as yet unmilked; for the family had on this day exceeded their usual sleep, and we have no heart otherwise than to excuse them, for they too had a week of toil.

In the tranquility of this Sabbath morning thought steals unawares across the deep blue sea to the quiet glen among the hills wherein they first drew breath, to the little parish church around which lies their kindred dust, and in fancy they hear the little bell from under its hooded tower in echoes quivering through the stillness of the liquid air. Soberly as

becomes the occasion we go out into the open day to
have communion with nature and nature's God.   As
we walk over the unlogged ground, a whippoorwill
starts at our feet and drops down only a few feet
away.   And just under our eye, in no nest, but lying
on the dry leaves are her two eggs; our outstretched
hand is nearly over her, so strong is her affection,
and we think of a greater love.   " Often would I have
gathered thy children together, even as a hen gather-
eth her chickens under her wings and ye would not."

Seated on a fallen tree covered over with moss we
open our Bible, but our thoughts wander, our senses
are absorbed in this sanctuary of nature and with its
beautiful ritual.   In the religious light of the forest,
trees hoary and twisted with the strain of centuries
are seen venerable in their scars, and over the sun-lit
spaces, where shadows flicker and waver, are the
pure white lilies lifted on their slender stems, and the
sweet little snow-drop, to us the dearest of flowers,
though we love them all.   And from under that mossy
bank there gushes a spring of pure water, where the
deer slaked its thirst and the Indian had glassed his
bronzed skin as he stooped to drink.   Voices blend
with the silence and would seem to be part of it; the
subdued tapping of the woodpecker on the decayed
trunk; the partridge drumming through the mist
away down among the ferns; with whisperings mani-
fold, real or imaginary, for everything is expressive,
the air is burdened with thought, nature worships
and we too are lifted in adoration.

And thought wanders to the silent centuries of the
past, with their great burdens of secrecy, and in our
abstraction we think of sunsets in the illimitable
woods; of silent nights and mornings that were washed
in dew; of dreary winters full of solitary wail; of trees
that grew old and venerable and died, and sank amid

their fellows unnoticed and unlamented; of storms that burst from breathless calms, slivering aged trunks that had resisted the strain of centuries, strewing their paths through the forest. And the great scars were healed in the lapse of years, and the records were enfolded in the past. And the shifting drama moved on, directed by Omnipotence—the Everlasting God to whom its beginning in the distant twilight is even, but as a thing of yesterday. But we return, and as we enter beneath our humble roof, we feel that God may be, and is worshipped in other then in temples made with hands. After dinner a few quietly fall asleep, and we do not wonder. And the old clock, the dear familiar clock, that crossed the sea, and which had ticked out the dying moments of a mother's life, is ticking now and very audibly in the stillness. From a shelf nailed to the wall we take a book, we have but a few, and these chiefly religious, which assists in directing our thoughts, yet to little purpose, from a drowsiness that is stealing over us. And the hallowed hours flit past! Then the cows have to be brought in from the woods for the sun is getting low. As we go out on that errand, a deer starts from the edge of the clearance, of which little notice is taken. We listen! and fancy we hear a faint tinkling of the bell afar off, which we follow, and the cows are found and driven home in the dusk of the evening winding their way through the forest.

After the custom of our fathers, we meet in the evening for worship. On a plain deal table the Bible is opened, and a psalm is sung in our own simple way, laden with memories. It may be that the XXIII Psalm, than which there is nothing sweeter—indeed the sweetest of pastorals—is read, then the priest-like father as we kneel to engage in prayer, for the time, puts out the light; and thus in the darkness our de-

sires are lifted to Him who was with us on the stormy
sea, who led us through many dangers to a haven of
habitation, spread for us a table in the wilderness,
and hath made us glad in this land of our adoption.
Our supplication is before Him, the little household
is soon at rest, and the midnight hours steal on, the
moon shedding her mellow light in at our little
window, and flooding with radiance the little stump-
studded clearance ringed in with the quiet woods.

A RAISING "BEE,"

In groupings where the "grog-boss" sat,
What zestful tales were told,
What news discussed of other lands,
Albeit rather old.

In the concessions, there is a raising "bee," and
we attend in the afternoon of the day, as a guest. It
is a busy scene; men are here by the score, as it is a
double log barn, and a few rounds have yet to go up
before it is finished. Basswood skids with the bark
taken off are leaning against the building all round,
on which the logs are being rolled up, a yoke of oxen
are hauling them in for that purpose. On the corners
men are dexterously balancing themselves, and with
axes are fitting them down into their places with
singular readiness and skill; others below are starting
them on to the skids with handspikes, following them
up with long crutches, a vigorous push given at every
"O heave," which rings in echoes to the waste.
And jokes are passed and the laugh raised, and
questions have been discussed all through the day as
opportunity was given, even politics were introduced,
and why not? Did not the settlers on the old con-
cession resolve into a company, and had two news-
papers direct from Toronto, which were passed for a
distance of two miles over the line, from one to the
other, thumbed and soiled as newspapers read in that
way only are; and in the same spirit of enterprise, an
Ohio grindstone was procured at a very early date,
and placed in the centre of that locality, which was
fitted up on a wet spongy day in winter, and axes
ground on it till away into the night; nor were they
satisfied, till a Yankee fanning mill was placed on the

line in the same way. But we are digressing. Seated at a short distance, under a few boards, which shield him from the weather, is the "grog boss"—an indispensable character, whose qualities usually commend him to the office, so that his installation in the morning was little otherwise, than a matter of form. As a rule, he is a man with a limp or other physical infirmity, or possibly advanced in years, but he is invariably a man of character, being entrusted with large discretionary powers. His duty is to keep a restraining hand on the whiskey, a responsible, though not an onerous position, and as a rule, he exercises his authority in a way which gives an agreeable latitude to the subject, while the interests of sobriety are fairly conserved. Yet to any rule there is an exception, but this in confidence. A log is being rolled up the skids, when a cry comes from the cornermen, "hold on," and for a time the log is held in position. A little man is seen, even now we see him, with his bright, honest face, a perfect index to his character, dressed in what had been a coat of good quality, but to give it better adaptability to every day work, is cut across a few inches under that line, where the jacket ceases to be, and the coat begins. In some way or other, he forgets responsibility, his feet begin to move and strike out into dancing attitudes, which ultimately step off brisk and lively; absorbed in himself, his handspike drops connection with the log, as he bends to the exercise, with his eyes downward, and as his skirts are less subject to the law of gravitation, than they otherwise would be, their action is free and unimpeded, and quivers with wondrous rapidity. It was unfortunate, though the log went up all the same. Our little friend was a good man, and has been for many a year, as we believe, in a better world.

As the building gets higher, the lifts are heavier, and strength is put to the test. And so it is as the last rays of the setting sun die in the encircling woods, while the evening dusk is gathering in around us, and the building stands nearly completed, the plates only to be raised; but which are invariably the heaviest lifts, and nerves have to be braced for the occasion, before grappling with them, a few wait on their friend by the sheltering boards, but who has now withdrawn from his retreat in the cool of the evening. Each one feels an individual responsibility, and so the plates are lifted to their respective places amid shouts and cheers. As customary one more enthusiastic than others, after much flourishing, and no little noise, flings a bottle of whisky from the top, yet for what purpose, we never rightly understood, then a hurrying to the ground clinging from one log to another. After the corner-men have inspected their work from below, they wait on the "grog-boss," who cheerfully attends to their wants, now that the work is finished and all are thankful, that it has been without accident.

Tea over, the majority leave, others wait, giving expression to their feeling in song and sentiment, after the tables are cleared, with mirth echoing among the rafters.

THE OX.

Sagacious ox, mute honest beast,
We long together wrought;
How wise thou wert! yet thou hast left
With all you ever thought.

A more useful beast than the ox in the early days
never existed. We do not forget that at times he was
said to be breachy, that he would lay down fences and
enter the grain fields—there are spots on the sun!
See him in the logging field, straining at the great
logs, till the bow creaks in the yoke, and the chain
snaps as if it were a dry withe, and when the last
brand from the burning is taken into the woods, leav-
ing the land cleared, drawing a clumsy harrow over
the rough uneven ground, scratching in seed among
the stumps; at times panting in the heat, his great
tongue lolling from his open mouth, his driver crying
to him in language often ambigious, and with fearful
threatening, even at times profane—thought his rarely
—and at the pitch of his voice, as if the poor beast
were not gifted with the sense of hearing. Or see
him under the starlight on a winter morning, the trees
cracking from the intense frost, with nothing better
on his stomach, than cold raw turnips and a bite of
dry hay, taking the road with a grist to the distant
mill on a big wooden sleigh, or bearing to the market,
the grain that he had trodden out on the threshing
floor with his own hoofs, only the week before, and
coming home in the darkness with icicles hanging
from his great jaws, and with a strange working in
his throat from the weariness of the day. He was
even at church on Sabbath with his big conveyance,
though we incline to think this by way of experiment,

and not often repeated. He was in everything, and
he had everything to do, and how patiently he bore
himself through it all, though often treated with
harshness, yet ever giving of himself the best, nor
resting till his work was completed. Occasionally a
worthy Dutchman was seen driving his big Pennsyl-
vania mares in from Waterloo, across the concessions
to the future city, though as yet the horse had scarcely
a solitary stable in our midst. Even after he had, it
was years before he settled quietly down to hard
work. He had aristocratic leanings; to do anything
so ignoble in his eyes as taking out manure, he
unequivocally refused, by violently rushing back-
wards, when progression was looked for. He held
that, as a burden bearer, he had not anywhere a place
in the economy of our concessions, and he insinuated
by his actions, that the ox was his debtor to the
extent of a living. It was extreme pressure that
brought him to act otherwise, and for long the patient
ox was held a reserve in the event of a strike, nor was
he parted with until his spirited ally submitted to dis-
cipline, and had confidence placed in him. Poor
beast, he had his physical troubles as we all have,
often sorely afflicted with hollow-horn in the spring
of the year, when his horns were bored with a gimlet
and stuffed with a vile drug, a poor substitute for that
which an empty stomach needs the most, though we
deny that this was ever prevalent on the old con-
cession.

His memory will perish, though we often think of
our horned friend—and they had great horns in those
days—as we saw him moving soberly up and down
and across the concessions on one errand or another,
seldom in a hurry, but always respectable and so
easily satisfied. To us it is very real. We feel now
as if standing in the evening dusk waiting his return,

and as we listen with strained ear, we think we catch a distant rumble of his wagon afar off, as it bumps over the corduroy leading through swamps; we have not been deceived as now the barred gate is being taken down at the end of the lane, and our big dog—said to be Dutch—with his loose, hollow bark, runs to meet him. Looming through the darkness he comes up to the door, where he is unyoked and turned into the little pasture to gather his meal of grass, where he rests for the night. He did a great work in his day, for which he was but poorly requited, and it is questionable if the concessions will ever rise to a true realization of his usefulness. In his history there is something regretful, touching and tender, and asks the tribute of a sigh. Moses like, he meekly led us in and out of the wilderness; but it was not for him to participate in the higher cultivation which should follow.

THE OLD BY-ROADS.

The old by-tracks, we knew them well,
　　They served a day of need.
Where did they lead to ; who could tell
　　To where they did not lead?

Yes, the old by-roads, or to be more correct tracks,
were cut, though not in every case, before the con-
cession lines were opened out. In those days traffic
objected to being confined to the straight, legalized
roads with their sharp, exacting corners where there
was no necessity for it and where a point could be
reached in a shorter and easier way and at little cost.
And there was an easy gracefulness in the sweeps and
curves of the old by-ways that was captivating to the
romantic mind as they deviated at their own sweet
will, now at the rear end of lots, then angling to the
front, again skirting the border of a swamp or swale,
then winding across the side of a hill, grazing the
edge of a little clearance filled with smoke, the tinkle
of the cow bell coming up from the hollows with the
deeper bass tones of other bells on the necks of the
working oxen. Nature was here in all its simplicity—
the ground-hog crouching by its burrow on the side
of a little hillock, the squirrel sitting with provoking
impertinence on the trunk of a fallen tree, the tapp-
ing of the woodpecker, with much that could easily
be mentioned, lent a charm to the old by-track that
was peculiar to itself. No stately equipage is seen to
disturb the serenity of the humble pedestrian. Duty
leads us to one of them this Summer afternoon as the
sun is near to its setting and we see something of the
traffic which passes over it. These girls whom we
are meeting have been to the village with eggs and

butter and are returning with empty baskets. They are really weary, we know they are, and they would never think of denying it, as they have had a long journey but they know nothing of ennui, nothing of that languor that comes of ease and luxury. Their lives are real, and when the individual life is merged into that of another, in the fuller relationship it will not be found lacking. And through among the trees an object is seen approaching us. It is a yoke of oxen with a wagon, winding along the side of a hill, the wheels slowly rising over the roots of the trees, then creaking down into the hollows, lifting over the cradle-knolls, lurching from one side to the other. With an earnest, plodding step the poor beasts come up to us, their noses nearly touching the ground, what sober faces, as if burdened with heavy responsibility. Weary and tired how gratefully they respond to the word "whoa." They are returning from the mill with a grist and with it a keg of whisky, which is nothing unusual, it is a necessity, at least so it was considered, for the wells as yet were not deep, consequently water was far from being satisfactory. It was the easiest thing in the world for the driver to draw the bung and give us a tasting—he would be pleased to do it, but it is unnecessary, though he may have occasion to draw before he gets to the end of his journey. Yet, do not associate him with low dissipation, rather follow him to his home on the concession and you will there learn that he leads an industrious life and that the bread that he eats is not the bread of idleness. There was little variety to the travel on the old by-tracks, no pleasure travel of any kind, if we except to the half-yearly Fair, which could never be neglected, though to buy and to sell and to get gain were scarcely the ends sought and obtained. It was a great social gathering to which the people came in from the clearances twice in the year to the Fair that was held in the village. It was truly an event and in which the social element was conspicious. Could it be otherwise? as four glasses for a "yorker" were offered by the various hostleries while heavier transactions were on easier terms, the effects of which were inspiring and though many would leave early in the day quietly, impressed with the sober realities of life, yet, with others, as the hours flew past, the interest increased, until unconsciously the darkness of night closed in ere they were aware, and with it the usual difficulties of starting for homes that were miles away.

Yet on every recurring Sabbath day, the little forest homes were represented in the village church. After the simple duties of the morning were attended to and worship over, some of the family at least, were off to church, through the woods, in the contemplative rest of the Sabbath morning; active girls, whose worth and beauty had captivated manly hearts, barefooted, their shoes and stockings in their hands, and putting them on after washing their feet in the creek that ran near to the village. And here in a little wooden church, a plain service is conducted, no choir, either surpliced or otherwise, but right under the genial minister is the precentor, who leads in the psalm—he leads Sabbath after Sabbath from a book that is bereft of one of its boards. Occasionally a snuff-box is seen wandering in and out among the pews, but skillfully guided until at rest in the pocket of its rightful owner, before the service is ended. If in the fall of the year, they are seen returning in groups, as the shadows lengthen, discussing a variety of topics, though very likely that venerable man with grey hair is deep into the mysteries of our religion. It is easily seen that he is familiar with his subject and is conscious of its gravity.

And thus through the quiet shades of the forest, they journeyed over the old by-paths, unseen by the world, often in difficulties, yet patient and hopeful, until the woods through which they led were but memories, indefinite and far off, even like to the wilderness of Kedar.

TROUT FISHING.

The speckled trout when we were boys,
That finned the shady streams,
And glanced above the sandy bars,
Are flashing through our dreams.

It is in the fall of the year, and the wind from the east has been sobbing over the stubbles all day, and now in the afternoon we are clearly in for a set rain. It is too wet for outside work, though all right for fishing. So we take our rod, a lithe little cedar, and with eager anticipation are soon at the swamp, near to the upper end of the old concession. It is still there, but "the glory is departed," its width and density are gone, and the tall pine trees, that lifted their tops against the sky, and were seen from afar, are away. Swiftly out from the cedars, the creek glides with a beautiful ripple across the highway, now larger from the late rains. We bait our hook as cunningly as we can, drop it into the water and only get a few unsatisfactory nibbles with which we have no patience. So we follow up to where there are fine mossy banks, drooping over and into the clear channel; we drop, and, quick as thought, there is one struggling in mid-air at the end of the line. This we string on to a switch, running the small end of it underneath the gills, and then we bring them out one after another, till we fancy their fears are awakened to a sense of danger and they are shy.

We go up a short way, no need to go far, as we know the place thoroughly, and as a haunt for trout it is perfect. Keeping back from the edge, we come to a moss covered log, half bedded in clear running water, and stealthily rising ourselves up we look over, and with breathless joy, see real speckled beauties, large as the heart could wish, swimming in and from under the mossy log, over the sand and gravel in trout-like majesty.   None of your latter day chubs, that have crept into our waters in a way only known to themselves, were there surging in manurial soakage from cultivated fields, but trout of·noble and ancient lineage, dating back from a time to which the landing of Jacques Cartier is but as yesterday.   Only think of them tracing the sinuosities of the stream beneath the gloom of the cedars, in the days of Abraham; or in the setting sun, rising to the insect brood and falling with a plash, on silence, that never answered to the voice of man.   Cautiously, we drop our line into the water, and at once get a vigorous bite, the line tightens, bracing ourselves to the occasion, we quickly lay him out on the wet leaves.   With a pecu-liar sensation, which can only be understood by experience, he is released from the hook.   We drop again, and again, we drop, and go up and down the creek and catch until we are perfectly satisfied, and then come away; for it is really wet.   Scarcely a solitary creature is seen, except occasionally a squirrel running as it were between showers from one place to another, or a crow sitting moodily high up on the leafless branch of a decayed trunk, or a bluejay wet and dripping, so we cut across the fields casting side glances on our string of trout, passing where ·the cattle are standing with drooping heads in the fence corners, the rain dripping from their sides and on to an open fireplace, that bids a kindly welcome.   Wet to the skin, the heat is enjoyable, and fires were fires then, when wood was had for the cutting.   With little delay the trout are laid in the frying pan, rolled in oatmeal and sizzling in the sweetest of butter.   In the delicious aroma which they emit, a savory fore-

taste is had of the riches to follow. The salivary
glands are in sympathy, so the mouth begins to water,
and with wistful eyes, we look on them getting crisp
and brown, and coloring to the perfect shade. Need
we say, how they were relished in the eating? It is
unnecessary, and only add that oysters, lobsters or
anything that lives and moves, and has its being, in
either salt water or fresh, are unworthy of being men-
tioned with them in the same breath.

THE NEW YEAR.

In viewless dress, through forest deep
The old year glides away,
We greet the new with brimming cup,
Our spirits fresh and gay;
We lead across the plain rough floor
In light subduing dusk,
The heart beats time to love and joy,
The feet to " Money Musk."

Away to the east, another year has been born into
the world. In a few hours, it will be gliding swiftly
up the snow-bound shores of the St. Lawrence,
muffled to the chin, seated in a cariole, or some other
conveyance of which we are uncertain, and is
expected to enter our concessions by twelve of the
clock. In New Germany, even now, guns are boom-
ing through the January thaw, so frequent in those
years, out from beyond the hill of Laubers, and the
creek that skirts the borders of that settlement, and
many a brimming cup of "lager" is drunk, and will
be in honor of its coming before day. True to the
minute it is with us; the old year quietly drops back
to be numbered with the years beyond the flood, the
new is welcomed with joy, and the German guns
redouble beyond the creek; youth, on whom the cares
of life sit lightly, are up and down the concessions,
with bottles primed with real "Old Allan," and enter
the homes along the line, we are besieged, and rise
out of bed, bottles are thrust on us and we drink,
though not with relish; for, oh! it is cold, and at
such an hour, but it is kindly and with the best
intentions. Then the jubilant tread, a measure to
give vent to their feelings, and though excitement
subsides with the dawn, yet a simmering is observable

all through the day, sustained by the reflection, that
—" It is, but ae day o' our lives, and wha wad grudge
though it were twa."

But the event of the season was a ball. It grew in
the early years into fame, waxing beyond the conces-
sions, and ultimately died after a long and brilliant
career, through its sheer popularity. On the evening
of the event, the youth and beauty of the settlement
came trooping in from the four roads, on foot, to the
quaint little schoolroom, swept and garnished for the
occasion; wraps were stowed into the teacher's desk
to the extent of its capacity, which was wonderful on
such a night. Need we tell of the floor and its allure-
ments, of music's exhiliarating strains drawn from the
sweetest of violins, of the joy of the hour, of the
abandon of the crowded floor, òf the hip hurrah! in
the morning hours, when footing to " double quick,"
and when the jaded dancers were in need of relaxa-
tion, how a song was asked for and given—a song of
their native land—or some other, not forgetting one
written expressly for the occasion, by a local Bard, in
which at the outset we were asked to

"Think upon Columbus, that man of worth and fame,
Who found out this great continent, that should have borne
     his name,"

with other couplets of equal or of superior merit.
And out under the starlight, a sugar kettle was
suspended over a huge fire, in which water was kept
boiling all through the night, for whiskey toddy was
served, and as needed, water was carried from a rivulet
in the vicinity. Boys, selected from the ranks of the
school room, were entrusted with this service, who in
exchange had a free run of the ball, with the right of
sharing in the ample stores that were provided. A
huge currant loaf had been compounded in the locality,
by a friend of the institution, and cheese and crackers

were in abundance, and O! how these eatables were relished by the privileged minors—the writer had a position on the staff.   Occasionally, we had the privilege of the floor, and what a moment of suspense that was as we stood in position, nervously waiting for the violin to sound the starting note, and then gliding off in the subdued, struggling light of tallow candles, in a whirl of excitement out and away into the mazy, wondrous intricacies of a Scotch reel.   A table was placed in a corner of the room on which the toddy was brewed by a practised hand, a man of sterling qualities, who in critical cases, or in the niceties of the art, conferred with others associated with him in the work—an interesting group—a study. And the winged hours flew past, and too swiftly, the company dispersing under cover of the night, leaving only a few who were seen in the steel grey of the morning like to the " thin red line " of Balaclava after the contest.

THE OLD LOG SCHOOL HOUSE.

Away far off through greying mists
    Of years that intervene,
The old log school house still exists
    With us a living scene;
And still to us the native woods
    Do fling their shadows grey
Across its low, flat roof and still
    We hear their organs play.

So the season quieted down the little school
room, useful in various ways, was given over to its
legitimate work and children were trooping in from
the four corners, as usual, to their accustomed
places.

In the weird light of that long-ago restless little
forms are seen fading away into obscurity, seated on
hard, backless benches around an open fire, their feet
buried in dust and ashes, with the virgin soil for a
hearth, their heads muddled and perplexed, often en-
gaged in pencil trading on the sly or in other ways of
peculiar and absorbing interest. And we realize the
ominous dread, which at times hung over the whole in-
fant mind, when a cloud was seen gathering at the desk,
and what relief was experienced should it disperse
and no one injured. Our presiding genius was prone
to extremes and was erratic, yet the little rustics saw
in him a man of rare acquirements, and in penman-
ship, either plain or ornamental, he excelled. Birds
on the wing would start on the page under his magic
pen reminding us of the lion as he rose in majesty
from the dust by creative fiat, as described by Milton.
In the course of years he abdicates and another lifts
the sceptre, but he sways it with a feeble hand.
Honest man ! see him as he stood, slovenly in his

attire and sorely in need of being better buttoned, his
head clearly indicating a Celtic origin, rough and
shaggy, his mind a perfect gallery of Hebrew bards
and prophets and full of the superstition that lingered
in his Highland glens in the days of his youth ! He
is seated at his desk, the heat is somewhat oppressive
and as it is just after dinner, he possibly feels a little
drowsy or he drops into an abstracted mood; not so
his sharp-witted pupils who take in the situation and
out from an open window hop one after another till
only two are left and they not through fear but out of
respect of constituted authority. He runs flusteringly
to the door, but only to see them skipping round the
four corners like deer. Was there an awful reckon-
ing ? Not a bit of it; we cannot even think that the
sun went down on his wrath and more than likely it
was quite forgotten. Authority is committed to one
who is efficient and skilful, earnest, conscientious
and intellectual, and we grew and fame spoke of us
beyond the concessions and the latter years of our
little school house overshadowed the first.

In the winter of '37-8, when the land seathed with
discontent and the sword of revolt was drawn, what
politicians we were ! We ourselves were in with the
people and were stirred with indignation, eager to do
great things and longing for an opportunity, yet with
glib little tongues expressing our views with energy
on the absorbing topic of the day.

And at the close of the winter seasons youthful
graduates were ever leaving the seat of learning to
engage in the summer's work, or when the fallen
leaves and the beach-nuts were covered with snow,
in lifting the axe against the thick trees that were
still rank in the forest, or coming on when the woods
were getting thin and with others assisting in clearing
up the fragments that were left and participating in
the universal joy.

And we see lithe little figures flitting like shadows
in the setting suns of bygone years, striving to get
the last tag of each other as they part for the day at
the four corners, and we gaze on them till the heart
yearns and grows very tender. Alas! these are fast
fading memories seen like to islands that we are leav-
ing far behind on the stream which we navigate, with
fog settling over them.

A COMMUNION SEASON.

Concessioners in simple form,
  Within that little room,
Commemorate the death which paid
  Their penalty of doom;
And tender thoughts flit to the time
  When on yon Sabbath day
They drank of the communion cup
  With dear ones far away.

In the initial years roads leading through swamps
were rough in their corduroy, yet heedless of such
hindrance the minister of Christ met with the settler
in his home and was received with joy. And men—
worthy men, who were scarcely in the ranks of the
ministry, yet eager to tell of Jesus and his love, seek-
ing no reward for their labor, save an approving con-
science—English non-conformists singing the hymns
of Watts and Wesley came among us and many a
Sabbath afternoon in Summer was made glad by their
services. Yet the message preached, ingrained with
the stern theology of Calvin and with the old Scotch
psalms—crisp in flavor—not a few of our pioneers
enjoyed with exceptional relish.

And a scene yet lingers in the memory of a few—
alas ! how very few are they—who can recall that
beautiful day in Summer, with a blue sky bending
above the concessions, and on this day (it is the
Sabbath) there is that peculiar and absorbing repose,
that intensified sweetness which we ever associate
with the day of rest. Even our own little school
house with its low, flat roof has a more sober look than
on other days, and the play ground out before the
door, with the great elm tree lying in its entire length
as it fell from the axe dividing it from the woods, is

still and noiseless and groups of twos and threes
come up from the four corners into the little area be-
fore the door in the crescent embrace of the forest,
and exchange greetings with one another and breathe
the freshness as it is wafted out from under the tall
trees, and linger in the pure sunlight till the minister
comes. He has been lodging for a few days in the
settlement, pursuing his sacred office, for as yet the
little flock who this morning come in from their for-
est homes to worship have no settled shepherd of
their own, and the occasion is one of unusual solemn-
ity; the sacrament of the Lord's Supper is to be ob-
served and with him they enter in at the open door
and quietly take their seats. The communion table
is set in the length of the room with a white cloth
spread over it after the good old custom of our
fathers, and opposite to the windows looking into the
woods, which down the face of the hill slope into the
valley. Not a few have been looking forward to the
occasion with joy as he who would chiefly conduct
the service was one whose many virtues adorn the
true type of the minister. In the course of the ser-
vice he reads the text selected for the day and the
sermon begins. And now let us look at the speaker
for a moment as he addresses his audience. In
height he is above the average but spare and wiry,
dark and swarthy in complexion, grave and reveren-
tial, with a voice that is singularly impressive and
conveys much, while his words are earnest and
thoughtful and uttered as if by one who is himself
fully persuaded—a fit representative of men who in
days of fiery trial did fearlessly meet their scattered
flocks in the mist of the moorelands in the solitary
glens. He rivets the attention of his audience as he
discourses on the theme of redeeming love and the
wondrous sacrifice, with an eloquence we associate

with intensity. At the close of the sermon the few communicants rise to take their seats at the table while the hundred and third psalm is being sung. And how quietly they move! how suppressed the rustle of garments! how subdued! And the thankful psalm that had gushed from the heart of Israel's Shepherd King is sung to Coleshill, and what music that was! how holy and reverential! strains that were sung sweetly in Zion stealing over our senses with a sanctifying spell, like a holy dream, or like to the whisperings of the forest in the mist of an Indian summer morning.

And on this occasion, so rich and pure, surely he who leads has the richest of voices. We know nothing of alto and soprano, we are unable to classify in the appropriate language of harmony, but we know this, that the voice of that sweet singer has trilled on our inner chords for over half a century, and it trills today. And the audience is lifted in thought, to the awful spectacle of a crucified Saviour, to a sinless soul in agony for the sin of a ruined world, to that intensely solemn moment, when with drooping head, and with the death-pallor creeping over His bleeding brow, He uttered the undying words, "It is finished," and to the soul haunting sequel, for nature inanimate was moved, and as if conscious of the fearful tragedy, she veiled her blushes with the shadows of eclipse. And "this do in remembrance of me" is read, and as it was in yon "upper room" at Jerusalem, a blessing is asked on the bread and wine, and the consecrated memorials are passed from one to the other. Before we separate, the time honored Paraphrase is sung,

"O God of Bethel! by whose hand
  Thy people still are fed!
Who through this weary pilgramage
  Hast all our fathers led;

> Our vows, our prayers, we now present
> Before thy throne of grace;
> God of our fathers! be the God
> Of their succeeding race," etc.

It is sung with expression, and to an old plaintive air, which we absorb into our very soul, and the meeting is dismissed with the benediction.

Alas! where are they, who took the communion on that solemn occasion ?   Yes, where are they ? Memories of the quaint little school-room are dying with the oldest inhabitant, shadowy as the ghosts of Ossian, our little school mates are seated by the dusty hearth, feet that moved in the merry dance over the plain rough floor, sound through the intervening years in muffled beatings afar off, and the sweet singing of the Psalm on yon holy Sabbath day, languishes on the ear like to a voice that is dying among the distant hills.   Soon, the ploughshare will be driven over the spot where the little structure stood, with nothing visible that might tell of its simple tale, nothing to wrest it from oblivion, save that Time's hoary fingers may weave a legend from its memories, which the youth of a distant generation may rehearse in the long winter evenings.   The old log school house is now a ruin, its tenants are the owl and the bat, yet though it were only a memory, to us it is, as it will ever be, the Westminster of the concessions.

A FAMILY GATHERING.

See, some on foot come in the lane,
   And now we see a sleigh,
The dear old homestead greets them with
   A happy New Years Day—
The great red glowing back-log burns
   As in the days of old,
And we, in circle round the hearth
   A grand re-union hold.

Once again, we return to the incipient years of our concessions, for we delight to wander in the great reverential woods, to trace nature into her secret haunts, and meet with her face to face, as friend meets with friend; to see the deer start with raised head as it listens, and bounds lightly away in the shadowy light of the forest; the leaves of autumn falling around us and rustling beneath our feet; the sun laboring through the mist; to hear the sound of the settler's axe, and the crash of the falling tree, as it surges back on the centuries; to be with the pioneer as he builds his little shanty under the shadow of the tall trees, roofed over with basswood scoops, with its one little window on the back, to see the smoke curling in wreaths from its clay built chimney, curling up into the pure air, losing itself in the sun-lit atmosphere. In the lapse of the seasons, a stack of grain—one solitary stack—looms on the eye, gladdening his heart, for it is bread from the virgin soil, won by his own toil-roughened hands, an earnest of the fuller harvest to follow, after years of weary toil, when the stacks would stand thick behind the double log barn filled to its utmost capacity and overflowing.

And we picture the little clearance, widening year by year, and the larger field ripening to the harvest, and the weather-blanched settler wielding his cradle among the stumps, drenched in sweat, through harvests of joy and again of sorrow, when rust was on the wheat, as it rose in clouds at every stroke, filling the mind with uneasy forebodings of cheerless days in winter, of how ends would be brought to meet, yet hindrances were bravely met, the way open- ing out before him, the gloom giving place to a brighter sky, and hope led on. And nature had pleasant aspects. Lambs in innocence skipped round the stumps in the warm spring weather, for flocks were introduced, though the wolves made havoc, and the rooster crowed from his own manure heap, through foul and fair, then, the dignity of labor was recognized. The sensible girl of the concession, in her neat straw hat, got up in the locality, and shoes that covered an adequate width of virgin soil, on warm summer mornings, footed it through the woods with baskets of eggs and butter; yes, footed it to the embryo city and thought nothing of it. And grey haired matrons in Leghorn bonnets, that had seen years of service under other suns, with great project- ing fronts, giving to the face a far *ben* look, though of useful design, took the road with baskets, and though of slower foot than their youthful compeers, ever kept a steady eye on the object to be attained.

And surely wisdom directed the settling of our con- cessions, so agreeably was it arranged. Strangers fitted into their places side by side, as if they had been specially prepared; adjusted themselves to the peculiarities of one another—for peculiarities there were, and it may be that by virtue of them, a more perfect whole was the result. Social grades and dis- tinctions had scarcely a place; Jack was as good as his master, if it could be said that a master was there, and worth was an appreciated quality. The sympathy of a common brotherhood was felt and practised. Of silver and gold they had none, yet of such as they had, they were ready to give—a helping hand in the day of need was seldom wanting. If snow lingered long in the concessions, and the spring was cold and backward, and work pressing, in places where the plough was unequal to the task, a kindly neighbor was ever ready to assist. Or, if through misfortune or otherwise the little logging field lay on into the summer untouched, a ''bee'' was suggested, and on the morning of the day, ready handed men came iu from the different lines with their sober-faced oxen, the chains rattling at their yokes, and they hitched on and the work was done up, and a kindly hand in harvest was ever given with a spontaneity which in later years is rarely witnessed, and ties tenderer than the ties of friendship were formed. Youth was ever meeting with the beauty of the concessions, growing up in quiet and sequestred homes, opening out into the charms of womanhood beneath the parental roof, beautiful as the lilies that gem the woods in early summer, and just as pure; and eyes spake to eyes in soft and tender glances, and words of hidden meaning were uttered, and warm impressible hearts were touched and attachments formed; and youth thinking to be unobserved, yet seen all the while by keen suspicious eyes, steals out at the five barred gate in the dusk slipping up the concession, where she, who has won his affections, is in expectancy, urging the lingering hour that has been uppermost in her mind all day; and now when everything is still, quietly they meet, and have lover's talk on into the night. Oblivious of the passing hour, and so it happens as he returns stealthily down the concession, he is seen by an early riser as the voice of the whippoorwill is hushed on the approach of day.

Or, when the cows were milked and the sweet frothy treasure placed nicely in the cellar and supper over, on a fine summer evening with the sun sinking in the western woods, the active girl of the locality would take her hat and stroll quietly out to the gate at the end of the lane, and, as it happened, one who had room in her little world, somewhere in the region of the heart, came up the concession at the time, and, as proper they spoke with each other at the bars, and may even have lingered longer than they thought, but could there be anything more natural ? We often in life meet with each other in that way, as ships cross each other's paths in the moonlight out on the lonely sea. Call it accidental if you wish, we prefer the recognition of a higher ordering. But ah! these gates five barred, or otherwise, are clustered with associations; to what tender partings they have been witnesses; what words of pathos have been uttered, and what vows have been breathed beside them in the evening dusks of their history. Yes, and in the ruddy light of the sugar camp, under the silent stars, in the wet spungy forest, the fierce light beating the darkness back in among the tall trees, the oft repeated tale of love was told, and warm hearts throbbed and tingled to its tender accents as they ever did. And after many months and in the fullness of time, the marriage banns were published in the little wooden church in the village, from the precentor's desk to attentive hearers, to ripple crispingly over the youthful mind of the locality to some extent prepared for its reception; for it was observed that a little clearance had been made across on the other concession, and a house had been built—a log house—for the shanty was now giving way to the better dwelling. And in the efflux of days, the day of the wedding came on—the auspicious day—and friends gathered in, cheerful and radiant, and the minster was on time, and the happy twain were made one for life's battle. And kind wishes were expressed, though wedding gifts were neither costly nor numerous. If there were any, it was as a token of sincere friendship, expressing soul and sentiment, it might even tell of sacrifice and kept sacred through life.

And the dinner was eaten, as all wedding dinners are, and a glass of whiskey toddy was brewed and drank, in which the minister joined, and it may have been repeated, taste was not as yet educated to the use of water pure and simple on such an occasion; more especially as the keg had been recently filled, a bag or two of shrunken wheat and chess having been taken to the distillery and exchanged for the potent beverage. And thus seated at an open hearth, where burned a huge back-log, the hours slipped past almost unnoticed on into the night, when the new married pair, were escorted across the concession to their future home in the woods.

With increased responsibility, they entered on the duties of a fuller life. And their work was beside them, which was neither light nor easy, as had oft been testified; yet through shade and shine the siren voice of hope was ever whispering—press on! Of how the little clearance widened into fruitful fields we need not rehearse, the story has been often told; but it grew dear to them, as the result of their own toil and industry, and little incidents were occuring daily, that were binding them to it closely with strong and tender links. Yet the old homestead had ever its peculiar attractions, and on the first day of the year they were there, absorbing heat from the great glowing back-log, and eating dinner at the old familiar table as on other days. And the wisdom of age, stooped to the lispings and prattle of children in easy relaxation, and everybody was pleased. And the cattle in whose veins the blue blood of the short horns, as yet scarcely ran, had to be seen and inspected in the warm, snug stable, and there was so much to be said and seen, that the time seemed all too short.

And then there was a year, when she who was the light of the home, an affectionate mother was missing. She who was watchful of their every interest, a ministering angel in every season of sorrow, whose very weakness was strength, uniting all into one loving union, was no longer with them on earth—'' for God took her.'' And another new year came round, and the widowed father met them as usual, though it may have been that his hand had a warmer expression than in other years. Yet the name of the missing one was rarely uttered. And there was no need. It was seen that she was the uppermost thought in the mind of all, and everywhere her absence was severely felt; it was wisely left for the salve of years to close up the sensitive wound, though the scar would be ever there. In the course of events, it was too evident that the old homestead would have to be parted with, through no misfortune, or blame on the part of any, but providential causes were making it inconvenient to retain. It was sad to think of leaving the old homestead—a home, which their own hands had hewn out from the deep forest, with all which that expresses, and their very name to perish with it, even as a memory. It was woven into their affections in a thousand ways—every field had its history, every scene its story; with the joys and sorrows of years it was associated, it was sanctified by a mother's death and a mother's blessing, children were born into the world, and here their years of innocence were spent, and the dear familiar hearth, where youth had seen visions in the great wooden fires, was clustered with memories that would never die. It was sad, even to the writer it is saddening, and for the time we will draw a veil over the closing scene.

OUR FATHERS, WHERE ARE THEY ?

Our pioneers, who in the woods
The little shanty reared,
Our widowed hearts will nevermore
Be with their presence cheered,
No, nevermore their voices hear
In sadness or in glee,
All still in death, and only now
A fading memory.

Only a few suns had risen on us as a resident of
the concession, when under a kindly escort, we were
guided through the woods to the brow of a hill, which
slopes into a deep and wide spread valley. On the
other side, and across a sluggish creek, was the great
suggestive forest, unbroken, and skirting the horizon.
Close under our eye was a dwelling house, over which
the shadow of death had lately passed. With a keen
scent death had tracked his prey into our midst, "the
dark Huntsman" was abroad, and had sent an arrow
for the first time in the history of our concession on
its fatal errand. A man, venerable in his three score
years and ten, was stricken, and across an unlogged
field men, accompanied by others, are carrying his
remains down the face of the hill to their burial.
And there, low down in the valley, on a little sandy
knoll, near to the margin of the stream that stole
quietly through the uncleared woods, he was laid at
rest, on a beautiful day in June, after his many years
of useful industry in his native land. He saw only
one round of the seasons in his adopted country, and
then a grave in its virgin soil near to the old conces-
sion road which was then uncut, yet far from the
murmur of his native Tweed and the land of the
heather.

Again we are in a little room where lies a dying mother. A few sympathising friends are with the husband in his hour of sorrow. That worthy neighbor, with grey hair and seasonable words, and sweet experience at the mercy seat, is very acceptable at such a time; and woman with her soothing, gentle hand is here, to do any little service that may be required, though little can be done. And as expected after the turn of the night, watchful eyes observe a change passing over the thin, pale face, the last moment is at hand. The candle burnt nearly to the socket, sheds but a feeble light athwart the little room, and words are spoken only in whispers. It is the stillness of death; yet in fancy we hear as if it were the dull stroke of an oar through the darkness, the dying mother hears a voice calling her away, the pale boatman is at hand, and she, trustfully leaning on the arm of Him, who is whispering, ''when thou passest through the waters, I will be with thee,'' is borne out on the dark and troubled waste to a happier shore. And the room seems as if it were emptied, and had become a great void. Then the father bends over the children that are in bed, and with a husky voice, he tells them that their mother is dead! Dead! They scarcely realize its significance, yet feel as if a dark cloud had crept over them. And the day of the funeral drew on with more than Sabbath solemnity. Nature was in sympathy with the occasion. A dull raw November atmosphere veiled the sky, and hoar frost was on the ground. The last look was taken of the cold marble visage of her, who was dearly loved, the plain deal coffin lid was closed on the precious dust and placed on a wagon, brought to the door and driven away to the village burying ground, followed on foot by neighbors, angling through the forest.

So death was with us! Now lurking stealthily and shooting his secret arrow, or daringly bold singling out his victim; as the ceaseless moaning of the sea, the wail of sorrow was ever heard in one home or another, and the funeral train was ever leaving the desolated hearth for the hungry grave, that never says there is enough. Now it was " Rachel weeping for her children," for that happy faced boy, barefooted and bonnetless, who ran all day, at night weary and asleep, to be washed and put to bed, giving his mother a great deal of work, and yet the doing of it was to her meat and drink; or his gentler sister, artless and winning, who was woven into her sweetest dreams, was taken away; or man stricken in the midst of his days, "his purposes broken off, even the thoughts of his heart." And now by the open grave of one who was very dear to us, we stand with bowed head. With his tastes and habits we were familiar, even in his inner life he was known to us, as we saw him going out and in for over half a century; he grew to be part of ourselves and we esteemed him as a true friend. An old pioneer, he saw the place grow up around him, yet he never forgot the scenes of his youth, and to have had his ear filled with the murmur of his native streams, would have given him joy. But this was not to be. So he died, amid the scenes, which his own hands had helped to create, "brought to his grave in a full age, like as a shock of corn cometh in in his season."

And we see another, as if it were yesterday, we pass the fields that he had cleared from the deep forest, but he is not there; often the rising and the setting sun saw him at his daily toil. In the rest that he had won by industry, he visited his native land, but it spoke to him with a voice of sadness tinged with regret, and there were ghosts on the hills, so he turned from its shores to live on the fields, which his own hands had won and to die among his old friends.

And woman gentle and loving, died, and was buried. She who was so true and heroic, so ingenious in her resources, making the most of everything, where it was so much needed, so sympathetic, easing the restless pillow, soothing the fevered brow, making glad with the sunshine of her presence and never thought she was doing anything.

And thus one after another of the old pioneers passed away, the place that knew them so well was to know them no more for ever. And we are never more to see them as we saw them on earth; never more on earth to hear the sound of their voices, others fill the places which they had occupied and we are solitary, even in our own concessions. And time is bearing us onward, bearing us away from the years of the great glowing back-log, and our yearnings are unsatisfied.

LYRICS

GLIMPSES OF THE YEARS THAT ARE GONE.

To scenes of our youth we in fancy stray,
To the early years we would homage pay—
To years when the axe through the forest rang,
Ere the free winds over the meadows sang,
Illumined as if by enchantment's art,
The scenes they evoke on our vision start,
And voices are whispering soft and low,
To us from the years that were long ago.

A wee shanty basks in the setting sun,
In a rough little field from the forest won;
And pumpkins lie thick, with the trailing vine,
As we sit in the dusk of dear lang syne;
And the night shades deepen and all is still,
Save woods that re-echo the whippoorwill,
Or, the lonely owl that doth vigil keep,
Away in the swamp 'mong the cedars deep.

And we brush the dew, ere the level ray
Of the sun hath lustred the leafy way,
As we hunt the cows in the grand old wood,
Where secrets kept through the centuries brood,
To their tinkling bell we are onward led;
While the deer starts up from her dewy bed,
And the leaves whisper back to the winds low sigh,
And the partridge drums in the thicket nigh.

And now through the trees like a flickering lamp
We catch a gleam of the sugar camp,
And we sit by its side, and we muse and dream
Through the quiet hours, see the rising steam

From the boiling sap, hear its bubbling sound;
While the weird light flares on the forest round,
And we gaze on the sparks as they upward fly,
And clouds that are raking the midnight sky.

And we see far off through the winking heat,
A wee harvest waiting of ripened wheat;
And the reaper appears with his sickle keen,
And a stack looms up on the rustic scene,
And his heart is rejoiced as he sees it stand—
It is Eshcol grapes from the promised land;
And we feel though remote, how the heart was cheered
With bread from the soil which our hands had cleared.

And we see a house, through a winter's night,
In a field thick with stumps, snow-capped and white,
And we lift the latch on a scene of mirth,
Where the yule-log burns on the open hearth;
And the old songs are sung to the heart's warm beat,
And the rough floor resounds to the dancer's feet,
Unheeded the clock with its warning hand,
Though swift hours are winging like drifting sand.

And through the mist of ingathered years
A rough log structure in vision appears,
Low roofed and flat; here worshippers stand
With uncovered head, sun-browned and tanned,
In homespun clad, while their voices raise
The rugged psalm or the hymn of praise,
To an old time tune, and with less of art,
Than the fervent ring of a trusting heart.

From the early days we are gliding fast,
We are speeding away from the cherished past;
We are speeding away from the scenes of yore,
From the early days which return no more;
And yet through the years how a voice will start
To move on the chords of the melting heart;
Scenes gleam through the mist we never forget,
We cling to them ever with tender regret.

THE CUTTING OF THE FIRST TREE IN GUELPH.

On the 23rd of April, 1827—a wet showery day—in the evening after the sun had set, the first tree was cut in the forest, where Guelph now stands. John Galt in his autobiography says:—"Having been shown the site selected for the town, a large maple tree was chosen, on which, taking an axe from one of the woodmen, I struck the first stroke. To me, at least, the moment was impressive and the silence of the woods, that echoed to the sound, was as to the sigh of the solemn genius of the wilderness departing for ever. The Dr. (Dr. Dunlop) followed me; then, if I recollect correctly, Mr. Prior and the woodmen finished the work. The tree fell with a crash of accumulated thunder, as if ancient nature were alarmed at the entrance of social man, into her innocent solitudes with his sorrows, his follies, and his crimes. I do not suppose the sublimity of the occasion.was unfelt by the others, for I noticed that after the tree fell, there was a funereal pause, as when a coffin is lowered into the grave; it was however, of short duration, for the doctor pulled a flask of whisky from his bosom, and we drank prosperity to the city of Guelph."

> The sun had set—an April day,
> Was blending with an evening gray,
> That softly through the forest crept;
> The stillness of the centuries slept
> In wild retreats, and undisturbed
> The native haunts of beast and bird.

A lifted axe! an era's doom,
Is knelling through the forest gloom,
And when at length, with crash and swell,
The old historic maple fell,
 With echoes wide renewed,
All to the scene reflection gave,
As men beside an open grave
 Do feel in soul subdued.

Then by the fallen Titan's side,
That erst had towered in forest pride,
 They drank in happy mood,
Success to Guelph in *usquebae*,
Success on that her natal day,
 Amid the native wood.

In vestal light the forest threw,
Its branches to a stainless blue,
As wearily in from other lands,
The fathers came like pilgrim bands,
From travel soiled; and fain were they
At length their flagging steps to stay.

A change is seen, as witnessed where
The hammer shakes the burdened air,
In alters to the Triune raised,
As seemeth best where God is praised;
In streets that in perspective fade,
Where erst would fall the forest shade;
In living men, who recognize
The social link, the kindred ties
That make us one; in law whose reign
The arm of justice doth sustain.

'Tis well! yet, lured by fancy's ray
To native scenes, in thought we stray
And picture heights, now city crowned,
Superbly girt with forest round
That drowsily in the valleys stood;
While winding through the silent wood,
The nameless river rippling swept
Its rocky bed, where cedars crept
Close to its edge and heard the wash
Of waters through the silence lash.

In deep recesses, still and coy,
The leafy summer lurked in joy,
And breathed in zephyrs through the vast
Umbrageous wood that softly cast
Subduing shade, a stillness lay
In keeping with the chast'ned day—
A brooding calm that drowsily weighed
And hung o'er all; deer meekly strayed
And herbage cropped at ease, and graced
With antler crest their native waste.

And flushing sunsets bathed the woods
And sank in distant solitudes,
As west'ring up the valleys crept
The dusky night, that dark'ning kept,
Till fancy to the ear would bring
The beatings of her murky wing,
While through the gloom, the silent stars
Looked calmly twixt the rifted bars
Of ebon cloud; and up the height
The owl shot through the silent night—
Her weird too-whoo, that dying fell
On silence in the lonely dell,
And strewing winds with sullen moan,
Wailed through the forest vast and lone
As summer fled; on forests bare
Snow fluttered through the thin cold air,
While seasons rolled—yea, centuries swept
Unheeded past, no record kept
Except in rings concentric, grained
On aged trunks, that long had strained
And wrestled with the storm that lashed
Their twisted boughs, and thus were dashed,
    Till on the silence fell
The stalwart maple to its doom,
That surged in echoes through the gloom,
    An era's passing knell;
Adown the intervening years,
Our quick'ned fancy ever hears,
    A muffled crash and swell.

TO THE RIVER SPEED.

Pellucid Speed, the years are few
Since on your banks the forest threw
Unbroken shade, and wild flowers grew,
   And clamb'ring creepers swung;
The spell of ages round thee lay
Unbroken, as you rolled away
   All silent and unsung.
Fain would we trace each misty scene,
With all the changes that have been
Since first the foot of white man prest
Your wooded slopes, by nature dress'd;
To twilight ages, dim and pale,
When starting down the silent vale
   Your onward course began;
Here through a swampy level led,
There rippling down a stony bed,
   And murmuring as you ran.

Enshrouded all! no annals tell
How forests rose and forests fell,
And were renewed; then in decay
Again, like others, passed away.
Yet in those silent years remote,
What dim romantic visions float;
The Indian, then in haughty pride,
Would roam in freedom by your side,
Then by the plashy brink with care
He set the crafty trap and snare,
Or through the slumb'ring forest sped
With eagle-eye and stealthy tread,
In keen pursuit to track the game
That fell before his practised aim;
And when the dusky shades of night
Stole through the woods and dimm'd his sight,
He with unerring foot retraced
The trackless way—the leafy waste—
To where his lonely wigwam stood,
Its curling smoke seen through the wood.

His wants appeased, with hungry zest
He wearily laid him down to rest,
  While all around was still.
No sound fell on his slumb'ring ear,
Except your waters rippling near,
The hoot of owl, or quick and clear
  The notes of whippoorwill.

Those scenes are past, and scenes to-day,
Like former scenes, shall pass away,
And in the ever deep'ning shade
Of years to dim remembrance fade;
Still youth shall in your shallows lave,
And herds shall drink your limpid wave,
And moonbeams on your waters play
That then, as now, shall break away
  Adown the pleasant vale.
On, ever on, your current tends,
Until your happy murmur blends
  With nature's dying wail.

ON SEEING AN INDIAN MOTHER CONVEYING A CHILD'S COFFIN DOWN THE STREETS OF GUELPH.

Bleak clouds floated over the sky, dark arraying,
  Snow covered the street and the landscape away,
As a poor Indian mother was lonely conveying
  A coffin she drew on a rude-fashioned sleigh;
For death in the wigwam had turned it to sighing,
Papoose's pale dust in its rude shroud was lying,
The spirit beyond where yon dark clouds are flying,
  In realms where the red man shall wander for aye.

Cold looks were cast on the dark, swarthy mother,
  Few knew her state, nor were mindful to know;
None turned aside, with the heart of a brother,
  In this, her bereavement, their pity to show.
To a valley she went, where a streamlet was wending,
Where the smoke of the wigwam rose upward, ascending
Through the tall forest trees, that around it were bending,
  And sighing a dirge o'er the ashes below.

Away from the wigwam, papoosie conveying,
  They scooped out a grave from the snow-covered ground;
Then ere departing, their last tribute paying,
  They chanted the death song in weird strains around.
Alone in the forest papoosie lies sleeping,
Where the night-hawk and owl lonely vigil are keeping,
With the wild wintry wind through the tall branches sweeping,
  Far from the place where the tribesmen are found.

Yet, may that mother, in her heart-felt emotion,
  As she thinks of her babe, drop the sad silent tear;
Shade of a race, that from ocean to ocean,
  Must vanish till nought of a remnant appear!
Once majestic to roam through their dark forests waving,
The storm and the tempest alike proudly braving,
From the hand of the pale-face no boon humbly craving—
  Disdaining to plead, and a stranger to fear.

ON SEEING A STUFFED PELICAN IN A STATIONER'S WINDOW, GUELPH,

Bird of the wild and solitary place,
　Of sedgy lake and lonely ocean strand,
I, in my fancy, would thy wanderings trace
　Throughout your native land.

I would thee trace, on bold adventurous wing,
　O'er trackless wastes decending to the north,
Where sighing winds, the harbingers of spring,
　Scarce call the blossoms forth.

And didst thou swim that cold mysterious sea,
　Where venturous sailor oft has found a grave ?
Or had those shores no heritage for thee
　Where rolls the icy wave ?

You may have winged the far Pacific slopes
　And rivers traced, that from the mountains run,
Whose rocky line and battlemented tops
　Shut out the setting sun.

You may have eastward borne your level way,
　Crossed silent forests waving in the breeze,
Till on your vision broke the rising day
　From our ''unsalted seas.''

You may have winged the Mississippi's course,
　No place for you those busy homes of man,
Then turned to trace Missouri to its source
　And far Saskachewan.

A glorious realm, but man, a tyrant race,
　Your life would seek, your liberties restrain,
If wandering in some solitary place
　The hunter took his aim.

So thou wert slain, and stand'st ignobly now
　The idle gaze of every passer by;
Could life again reanimate, how thou
　Wouldst spurn their gaze and fly.

More meet that thou by Winnipeg hadst lain
　Calm on its shore to rest thy folded wing,
Where sea-like lies the trackless prairie plain
　Winds should thy requiem sing.

WILLIE BUCHANAN.

General Middleton, in his report to the Governor-General on the Northwest Rebellion in Canada, says:—I cannot conclude without mentioning a little bugler of the 90th Regiment, named William Buchanan, who made himself particularly useful in carrying ammunition to the right front, when the fire was very hot, with peculiar nonchalance, walking calmly about, crying: "Now, boys, who's for cartridges?"

'Tis at Batoche's ferry,
  In wrath the bullets fly,
And fierce the battle rages,
  And men like heroes die;
The virgin prairie's crimsoned
  With blood that's freely shed,
The crack of gun and rifle
  Is telling of the dead.

Yet, in the thick of danger,
  The battle fierce and hot,
Where bearded men are falling
  Beneath the rebel shot;
Amid the bullets whizzing,
  And hissing to destroy,
What form is that, with agile limb
And dauntless eye? Who knows of him?
  A little bugler boy.

The dew of youth is on him,
  A hero! is he not?
His bugle notes are silent,
  He bravely carries shot.
He carries to the rifles,
  And aye we hear him say,
"Now boys, who's for cartridges?"
  And thus throughout the day.

Death strikes with shaft unerring
Its victims in his sight,
Yet self-possessed and calmly
He mixes in the fight;
He treads the reddened prairie,
Unconscious in his heart,
That in his way, amid the fray,
He acts a hero's part.

But fame, though oft capricious,
In wooing, shy and coy,
From off her heights is watching
Her little bugler boy.

She looks on Willie kindly—
She smiles with truest joy,
As on her roll of honor,
She writes the gallant boy,
Around the letters weaving
In wavy, graceful fold,
The old Buchanan tartan,
That oft a tale hath told.

Extol the youthful hero!
Weave garlands fresh and gay!
Go write his deeds in story!
And sing the stirring lay!
And when we tell of bravery
And gallant deeds with joy,
Then proudly think of Willie,
Our little bugler boy.

SLEIGHING.

O, haste us away for our hearts are gay
  And Blanche at the door is neighing;
And champing the bit, while her eye is lit
  With joy as she thinks of sleighing,
  O, the thrill on our senses playing
As we glide away on a pleasant day
  With zest in the best of sleighing.

On the icy snow how smoothly we go,
  Or rock with a gentle swaying,
As we cross the hills and the frozen rills
  The fields and the homes surveying,
  Where herds in the yards are straying;
While the ring of the bells on the crisp air swells
  In time with the rhyme of sleighing.

In our joy we laugh and the pure air quaff,
  The burdens of life unweighing,
As the crisping air lifts the clouds of care
  That leave us with nought dismaying,
  And when day in the west is graying
We are home again in the finest vein
  With gleams in our dreams of sleighing.

THE HIDDEN FUTURE.

The night with dusky mantle has wrapt the mountain's breast,
The weary foot of labor has sought a place of rest,
Our little ones beside us, with hearts so light and gay,
In happy glee their feet have ran through all the busy day.

And now each little cherub form beside the table placed,
The eye with youthful pleasure beams, no care the brow has
    traced,
And as they talk with simple tongue, they paint a future day,
A happy scene, with cloudless sky—a landscape glad and gay.

They dream their little fancy dreams and count the weary
    years
Ere yet erect they proudly stand as men beside their peers,
Within their native vale to live, or seek a foreign strand,
And laurels win of wealth and fame, and all be good and
    grand.

We listen to their childish talk, and strange emotions rise,
For oh! how soon their visions bright may dim with cloudy
    skies,
And wand'ring in the tempter's way what ills may them betide;
Our hearts are sad, yet trust that One their feet may ever
    guide.

Our hopes are oft delusive on life's uncertain way,
The light that shines upon our path is given day by day;
We scarce would dare to seek a change, the heart as truth
    believes—
The veil that doth the future hide, the hand of mercy weaves.

"THE HEART KNOWETH ITS OWN BITTERNESS."

There is care that admits of relief,
  To a friend or acquaintance revealed;
Yet we all have our unbosomed grief
  Apart from all others concealed.

The heart doth its own sorrows know,
  It has those that are shared in by none;
There are depths in the valley of woe
  Where we walk in the darkness alone.

In life there are solitudes dire,
  When the soul sinks and would be released;
We have longing and restless desire,
  That always remains unappeased.

With care and disquiet oppressed,
  The soul in its cabin of clay,
Doth yearn for expansion and rest
  In the light of an infinite day.

FLOWERS.

The flowers, how beautiful the flowers,
  So pure, so chaste, so exquisitely rare;
Whether in gardens or in forest bowers,
  They scent the quiet air.

But yet to me wild flowers by wood and rill
  More winning are and dearer to the heart
Than those we culture with the greatest skill
  Of science and of art.

The coy retreat, where trees their shadows throw,
  And veil the fervor of the noontide ray,
I love to seek, to see the wild flowers grow
  Where foot doth seldom stray.

God's garden this, His hand with pencil rare
  Doth tint their leaves and trace their slender stems,
Each want supplies, He nurtures them with care,
  And guards these forest gems.

And thus they grow 'neath storm and changing skies,
  'Neath falling dews and suns with tempered ray,
Till Autumn winds proclaim that summer dies,
  Then meekly pass away.

MEMORIES.

Forms are in the distance fading,
   There are forms that never fade;
Time may cast its misty shading,
   Still we see them through the shade.

Years may roll with care engrossing,
   Yet unchanged our hearts remain;
How a thought the memory crossing
   Wakes the past to life again!

Voices that were lost in sadness
   Break as through the startled air,
Voices once so full of gladness;
   Fancy fills the vacant chair.

Friends with whom in life we parted,
   Whom in knowing fairer grew,
Rise before us open hearted,
   Loved and loving, ever true.

Little hands and tiny fingers
   Press upon our bended knee;
And the voice in echo lingers
   Once so happy full and free.

Treasured in the heart's recesses
   All our tender memories lie.
Shall they live? The soul expresses
   Hope that they will never die.

ON THE STREET.

SCENE—*King Street, Toronto.*

The little roughened legs were clad
  In stockings nature gives to all,
No shoes—no, neither good nor bad,
  And his thin form was far too small
To suit the vesture that he wore;
Beneath his little cap a store
  Of curled locks his brow had graced,
But all unkempt; with eager feet
  A newsboy thus aleartly paced
His way along the city street,
  With stock in trade he entrance gained
  A restaurant where fashion reigned.

Here at a sumptuous table sat
  A youth—in truth a city swell,
With taste aesthetic, what of that ?
  'Tis of his act that we would tell,
  Of how it to the boy befell,
  Of how the little fellow gazed
  With glad surprise as if amazed
And seemed bewitched, as waiting not
  This man so exquisitely fine,
  This votary at fashion's shrine,
Did order supper nice and hot,
  And had it served, a feast of joy,
  To this poor little hungry boy,
  Not in a nook on dirty delf
  But at the table with himself.

Set out the picture full in sight,
Surround it with the clearest light,
Such scenes our better thoughts employ,
A scene to give an angel joy,
As here this man of fashion sat
And right before him little Pat.